The Open University

Mathematics Foundation Course Unit 18

PROBABILITY AND STATISTICS II

Prepared by the Mathematics Foundation Course Team

Correspondence Text 18

The Open University Press

Open University courses provide a method of study for independent learners through an integrated teaching system including textual material, radio and television programmes and short residential courses. This text is one of a series that make up the correspondence element of the Mathematics Foundation Course.

The Open University's courses represent a new system of university level education. Much of the teaching material is still in a developmental stage. Courses and course materials are, therefore, kept continually under revision. It is intended to issue regular up-dating notes as and when the need arises, and new editions will be brought out when necessary.

Further information on Open University courses may be obtained from The Admissions Office, The Open University, P.O. Box 48, Bletchley, Buckinghamshire.

The Open University Press
Walton Hall, Bletchley, Bucks

First Published 1971
Copyright © 1971 The Open University

Printed in Great Britain by
J W Arrowsmith Ltd, Bristol 3

SBN 335 01017 2

Contents

		Page
	Objectives	iv
	Structural Diagram	v
	Glossary	vi
	Notation	viii
	Bibliography	viii
	Introduction	1
18.1	**Randomness and Probability**	1
18.1.1	Randomness	1
18.1.2	Probability	1
18.2	**Sample Spaces**	2
18.2.1	Sample Spaces	2
18.2.2	The Cartesian Product of Sample Spaces	5
18.2.3	The Urn Model	7
18.2.4	Permutations and Combinations	10
18.2.5	Summary	13
18.3	**Rules of Probability**	15
18.3.0	Introduction	15
18.3.1	Rule 1	15
18.3.2	Rule 2	15
18.3.3	Conditional Probability: Rule 3	21
18.3.4	Applications	22
18.3.5	Statistical Independence	25
18.4	**Ascribing Probabilities**	28
18.4.1	Relative Frequencies	28
18.4.2	Equally Likely Cases	29
18.4.3	Definitions of Probability and Randomness	29
18.4.4	Subjective Probability	33
18.5	**Summary**	34

Objectives

The principal objective of this unit is to refine the notions of probability and randomness.

After working through this unit you should be able to:

(i) use the language of sample spaces and elementary events, and the associated algebra of sets;
(ii) calculate the number of permutations and combinations of r objects selected from n objects;
(iii) state and explain the rules of probability;
(iv) calculate simple probabilities using the urn model with or without replacement;
(v) explain the meaning of conditional probability and use it in probability calculations;
(vi) explain the meaning of statistical independence;
(vii) give a definition of probability in terms of sample spaces;
(viii) give the corresponding definition of randomness.

Note

Before working through this correspondence text, make sure you have read the general introduction to the mathematics course in the Study Guide, as this explains the philosophy underlying the whole course. You should also be familiar with the section which explains how a text is constructed and the meanings attached to the stars and other symbols in the margin, as this will help you to find your way through the text.

Structural Diagram

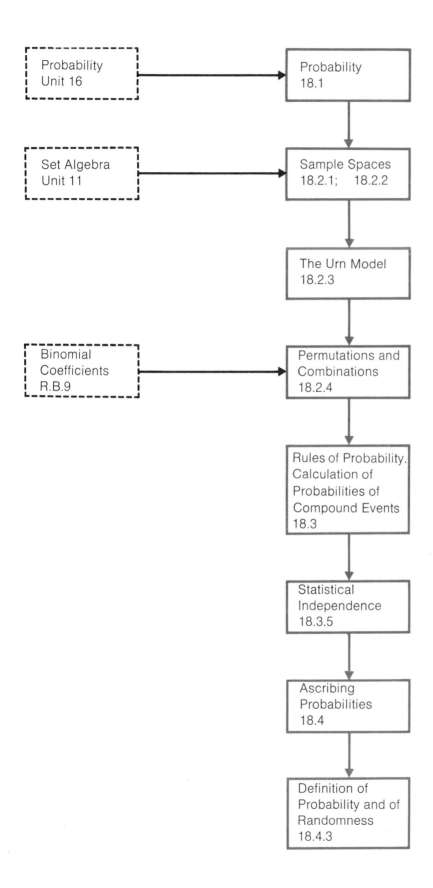

Glossary

Terms which are defined in this glossary are printed in CAPITALS.

COMBINATION	A COMBINATION is any selection of a set of r objects from a set of n objects.	11
COMPLEMENTARY EVENT	The COMPLEMENT of the EVENT A is the set-theoretic complement A', where the universal set is the SAMPLE SPACE of A.	3
CONDITIONAL PROBABILITY	The CONDITIONAL PROBABILITY of an EVENT A is the PROBABILITY of A given that an event B (say) has occurred.	21
DEPENDENCE (STATISTICAL)	EVENTS A and B having non-zero probabilities are STATISTICALLY DEPENDENT if they are not STATISTICALLY INDEPENDENT.	28
ELEMENTARY EVENT	An ELEMENTARY EVENT is an EVENT consisting of a single SAMPLE POINT.	2
EVENT	An EVENT is a subset of a SAMPLE SPACE.	2
EXCLUSIVE EVENTS	EVENTS A and B are EXCLUSIVE if $A \cap B$ is the empty set.	4
EXHAUSTIVE EVENTS	EVENTS $A_1, A_2, A_3, \ldots, A_n$ are EXHAUSTIVE if $$A_1 \cup A_2 \cup A_3 \cup \cdots \cup A_n = S,$$ where S is the SAMPLE SPACE.	20
INDEPENDENCE (STATISTICAL)	EVENTS A and B ARE STATISTICALLY INDEPENDENT if $$P(A \cap B) = P(A) \times P(B) \neq 0$$	28
PERMUTATION	A PERMUTATION is an ordered selection of r objects from a set of n objects.	10
PROBABILITY (mathematical definition)	If there is a function P, with domain the set of all EVENTS of a SAMPLE SPACE S, and codomain R, such that the images under P obey the RULES OF PROBABILITY, then these images are the PROBABILITIES of the corresponding events.	29
RANDOM NUMBERS	RANDOM NUMBERS between 0 and 9 are the recorded outcomes of a TRIAL which is RANDOM with respect to the SAMPLE SPACE $S = \{0, 1, 2, 3, 4, 5, 6, 7, 8, 9\}$ where each element of S has probability $\frac{1}{10}$. Random numbers between 0 and 99 are defined similarly.	32
RANDOM SEQUENCE	A RANDOM SEQUENCE is a sequence of outcomes of a RANDOM TRIAL.	1, 31
RANDOM TRIAL	Given a finite set S such that each element of S has a given number associated with it, where the numbers obey the rules of probability, a TRIAL is said to be RANDOM with respect to S and the associated numbers if it has SAMPLE SPACE S such that the given numbers are the PROBABILITIES of the corresponding ELEMENTARY EVENTS.	30
SAMPLE POINTS	SAMPLE POINTS are elements of a SAMPLE SPACE.	2
SAMPLE SPACE	A SAMPLE SPACE is the set of possible outcomes of a TRIAL.	2

Page

SELECTION WITH/ WITHOUT REPLACEMENT — In terms of the URN MODEL, SELECTION WITH/ WITHOUT REPLACEMENT is the selection of balls one at a time from an urn, where a selected ball is replaced/not replaced before the next ball is selected. — 7

SUBJECTIVE PROBABILITY — If a person has various degrees of belief or confidence (which may all be the same) in the various outcomes of some TRIAL, and quantifies these degrees by numbers satisfying the RULES OF PROBABILITY, then these numbers are the SUBJECTIVE PROBABILITIES of the person concerned. — 33

TREE DIAGRAM — A TREE DIAGRAM is a diagram (composed of branching lines) which shows the structure and the SAMPLE POINTS of the Cartesian product of two or more SAMPLE SPACES. — 5

TRIAL — A TRIAL is an experiment whose outcome need not be the same every time it is repeated. — 2

URN MODEL — If the outcomes and PROBABILITIES of a TRIAL are analogous to the outcomes and probabilities associated with the selection of balls from an urn, the latter situation is called the URN MODEL. — 7

Notation

Page

The symbols are presented in the order in which they appear in the text.

^nP_r	The number of ordered selections of r objects from a set of n objects.	10
	$$^nP_r = \frac{n!}{(n-r)!}$$	12
$n!$	n factorial, that is, $n \times (n-1) \times \cdots \times 3 \times 2 \times 1$; 0! is defined to be 1.	12
^nC_r or $\binom{n}{r}$	The number of combinations of r objects from a set of n objects. $$^nC_r = \frac{n!}{(n-r)!r!}$$	13
$P(A)$	The probability of the event A.	15
$P(B/A)$	The conditional probability of event B, the condition being that event A has already occurred.	21

Bibliography

F. Mosteller, R. E. K. Rourke and G. B. Thomas, *Probability with Statistical Applications* 2nd ed. (Addison-Wesley 1961).

The first 170 pages of this book follow *Unit 18* closely. The last half of the book deals with issues and techniques beyond the scope of this course. However, if you intend to continue your study of statistics after the foundation course, this is certainly a good book to have.

S. N. Collings, *Theoretical Statistics: Basic Ideas* (Macdonald, to be published in late 1971).

This book develops the subject from scratch, and its earlier chapters are very similar to this correspondence text. It develops the subject beyond the foundation course syllabus in very much the same style.

18.0 INTRODUCTION

In *Unit 16, Probability and Statistics I* we met sequences of 0's and 1's whose terms were unpredictable and patternless. Such sequences were said to be *random*, though no formal definition was offered.

We found it difficult to frame formal definitions of *randomness* and *probability* without becoming involved in a circular situation. The main purpose of this unit is to break this circularity as best we can; we do this by switching to an axiomatic approach to probability. We make sure, however, that the behaviour of probability as specified by the axioms — or *rules* as we shall call them — is such that it corresponds to the intuitive properties of relative frequencies. Having established probability, we shall then be able to offer a formal definition of randomness.

In the course of this main thread, we shall introduce new notions such as *sample spaces* and *the urn model*; also special mention will be made of *conditional probability* and of *statistical independence*.

18.1 RANDOMNESS AND PROBABILITY

18.1.1 Randomness

Let us begin by reviewing the situation, and then carry the argument a little further. In *Unit 16* you carried out a card guessing experiment producing a sequence of 0's and 1's. We have already referred to this sequence as being a *random* sequence, and we have identified this word with *lack of pattern* and *unpredictability*. It is difficult to put these negative attributes into positive terms. We are tempted to say that the sequence is random because at any stage — and whatever the form of the sequence so far — there is the same uncertainty as always about there being a 1 next time. In other words, the probability of a 1 at any stage is always the same, irrespective of the results so far. The trouble with this as a definition of *randomness* is that it presupposes a definition of *probability*. Perhaps a better description of a random sequence is one for which the only way to specify the complete sequence is to write down every term — there is no formula for and no way of predicting any specified term in the sequence — but this again has difficulties. We observed that the relative frequency of 1's in our random sequence showed signs of tending to a limit. It is easy to imagine sequences in which the relative frequency of 1's oscillates between small and large values. If this occurred in your experiment you would with some justification conclude that the order was not random.

18.1.2 Probability

Given a random sequence of 0's and 1's, we have accepted as an experimental fact that the relative frequency of 1's behaves as if it were tending to a limit. If the sequence could be continued indefinitely and the value of the limit obtained, that value would be the *probability* of a 1 in a single trial. The trouble with this description of *probability* is that it presupposes a definition of *randomness*.

We require the sequence to be random, for otherwise we have no reason to expect that the relative frequency of 1's will tend to a limit. But randomness is more deeply involved than this. It is easy to construct non-random sequences for which the relative frequency tends to a limit; for example:

$$0, 0, 1, 0, 0, 1, 0, 0, 1, 0, 0, 1, 0, 0, 1, \ldots$$

The limiting value of the relative frequency is $\frac{1}{3}$ in this case; but it would be nonsensical to suggest that the 100th figure has probability $\frac{1}{3}$ of being 1.

Randomness cannot be defined without a back-handed reference to probability, but probability is an attribute of random sequences. We must find some method of breaking this circularity.

If we cannot define probability precisely (in fact, in a restricted way, we do define probability in section 18.4.3), does this imply that we cannot carry the subject any further? No! We learn to use numbers long before (if ever) we have been told by analysts and mathematical logicians what numbers are;* this is because we can learn to combine things by rule (or axiom) even if we do not know precisely what they are.

In some ways our situation in probability is much the same. Given a die, we could argue about whether the probability of getting a 1 was $\frac{1}{6}$ and of getting a 2 was $\frac{1}{6}$, or even about what such statements mean. But whatever is meant, if the probability of $\frac{1}{6}$ is correct, then we would infer (by intuition, or whatever) that the probability of getting a 1 OR a 2 is $\frac{1}{6} + \frac{1}{6} = \frac{1}{3}$.

18.2 SAMPLE SPACES

18.2

18.2.1 Sample Spaces

18.2.1

As we cannot immediately escape from the circularity mentioned in the previous section, we begin the subject of probability again, and see how far we can get by by-passing trouble. The first thing to notice is that probabilities are concerned with outcomes of trials. Let us therefore consider outcomes as a separate topic on their own.

Main Text
* *

In the situations which we consider in this unit, a trial has a number of possible discrete outcomes. If we were throwing a die, the outcome would be one of the integers 1 to 6; if tossing a penny, the outcome would be a head or a tail. A set of all possible outcomes is known as a sample space. What exactly we mean by a "possible outcome" is left to intuition: thus we do not include the possibility of the penny landing on its edge. In some cases the determination of the sample space to be used can itself be a problem. The individual outcomes making up a sample space are known as sample points.

Definition 1
* * *

Definition 2
* * *

Given a trial having sample space S, any subset of S is called an event; a subset consisting of a single sample point is called an elementary event.† When we talk of an event E "occurring" we mean that one of the sample points which belongs to the set E occurs. For example, for the throw of a six-faced die,

Definition 3
* * *

  the sample space is $\{1, 2, 3, 4, 5, 6\}$;
  3 is an example of a sample point;
  $\{1, 2, 3\}$ is an example of an event,
  and $\{3\}$ is an example of an elementary event.

We would say that the event $\{1, 2, 3\}$ "occurs" if a 1 OR a 2 OR a 3 is thrown.

---

* We defined the number 1 in *Unit 17*!

† We shall not always distinguish between a sample point and the corresponding elementary event.

In the general situation there are four main cases:

(i) Since, mathematically, the empty set, $\emptyset$, is regarded as a subset of any set*, an event E can be the empty set.
In this case E contains no sample points, so it does not correspond to reality; therefore the event E is impossible.

(ii) E can consist of just one sample point. In this case E is an elementary event.

(iii) E can consist of some but not all sample points of S.

(iv) E can contain all the sample points of S, and so be identical with S.
In this case, the event E is certain to occur.

Example 1   Example 1

For the throwing of a six-faced die, the event corresponding to getting an even number is the set $\{2, 4, 6\}$; the event corresponding to getting a 7 is $\emptyset$.

Corresponding to the vocabulary we used for sets in *Unit 11, Logic I*, if the sample space is denoted by S, and A is any event (that is, any subset of S), the event consisting of all sample points in S which do not belong to A is called the complementary event to A. It is denoted by A', and is illustrated in the following diagram:

Main Text
* *

Definition 4
* * *

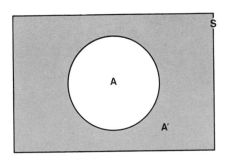

For example, if E is the event of getting an even number on throwing a die, the complementary event E' is the event of getting an odd number.

If A and B are two events, we can, of course, use them to define new events corresponding to the set operations. In particular, we can define the events corresponding to $A \cup B$, the union of A and B, and $A \cap B$, the intersection of A and B.

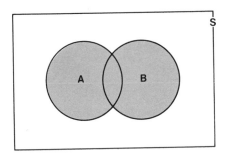

In the diagram the shaded area represents the event $A \cup B$.

---

* If A is any set, we say that $\emptyset$ is a subset of A, because $\emptyset$, being empty, has no element which does *not* belong to A.

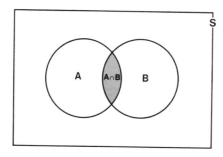

In the diagram the shaded area represents the event $A \cap B$.

For example, if S is the sample space corresponding to one throw of a die, and $A = \{1, 2\}$ and $B = \{2, 3\}$, then $A \cup B$ is the event $\{1, 2, 3\}$ and $A \cap B$ is the event $\{2\}$, which we sometimes write without the brackets.

If $A \cap B = \varnothing$, the empty set, then we say that the events A and B are exclusive.

Definition 5
* * *

In the following diagram, the events A and B are exclusive as they have no sample points in common.

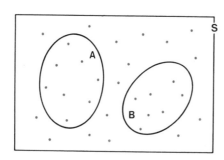

If events A, B, C are mutually exclusive (that is, any two of them are exclusive), then A and $B \cup C$ have no sample point in common; that is, A and $B \cup C$ are exclusive. Similarly if A, B, C, D are mutually exclusive, A and $B \cup C \cup D$ are exclusive; so also are $A \cup B$ and $C \cup D$. These results can obviously be extended.

Exercise 1

Exercise 1
(2 minutes)

(i) On a table there are five counters: one counter bears the letter A, another the letter B, and so on to the letter E. Two of the counters are removed and their letters noted. Describe the sample space representing all possible outcomes (all possible pairs of letters corresponding to the two counters removed). How many sample points does it contain? (In this question the order in which the two letters are selected is immaterial.)

(ii) In (i), let X be the event consisting of all pairs which contain the letter A. How many such pairs are there?

(iii) In (i), let Y be the event consisting of all pairs which contain the letter B. How many such pairs are there?

(iv) Describe the event $F = X \cap Y$; how many pairs does it contain?

(v) Describe the event $G = X \cup Y$; how many pairs does it contain?

(vi) Specify the event X', where X' is the event complementary to X. ■

18.2.2 The Cartesian Product of Sample Spaces

A sample space covers all the possible outcomes of a single trial, and each elementary event corresponds to one possible outcome. Often these outcomes are compounded of other outcomes. For instance, consider the sample space corresponding to the tossing of two coins one after the other. The set of all possible outcomes is

$$S = \{(H, H), (H, T), (T, H), (T, T)\},$$

where H stands for a head and T stands for a tail. Notice that because the coins are tossed one after the other, (T, H) is different from (H, T). Had the coins been tossed together, we might have preferred to use the sample space $\{HH, HT, TT\}$, of non-ordered pairs. The sample space for the tossing of one coin is

$$S_1 = \{H, T\},$$

and it can be seen that

$$S = S_1 \times S_1,$$

the Cartesian product of S_1 with itself. In general, if a trial consists of two or more parts, such that the sample space of the complete trial can be written as the Cartesian product of the sample spaces of the parts, we talk of a compound trial.

When dealing with compound experiments, it may be useful to draw a diagram such as the following, which is known as a tree diagram.

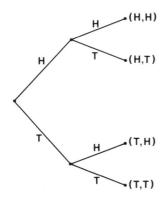

Tree diagram for $S_1 \times S_1$

It can be seen from the tree diagram that the number of sample points in the sample space $S_1 \times S_1$ is the square of the number of sample points in S_1. In general, if S_1 and S_2 are two sample spaces, and $S = S_1 \times S_2$ is the sample space corresponding to the compound experiment, then if $N(S)$ denotes the number of elementary events in the sample space S, we have

$$N(S_1 \times S_2) = N(S_1) \times N(S_2).$$

Example 1

Example 1

A trial consists of tossing three coins one after the other. Write down the sample space and draw the corresponding tree diagram. ■

Solution of Example 1

The sample space for tossing one coin is

$$S = \{H, T\}.$$

(continued on page 6)

Solution 18.2.1.1

Solution **18.2.1.1**

(i) The sample space, S, consists of the ten possible (non-ordered) pairs of non-identical letters from the letters A, B, C, D, E.

$$S = \{AB, AC, AD, AE, BC, BD, BE, CD, CE, DE\}.$$

(ii) There are four pairs making up the event X.

$$X = \{AB, AC, AD, AE\}.$$

(iii) Similarly, the number of sample points in Y is also four.

$$Y = \{AB, BC, BD, BE\}$$

(iv) The event $F = X \cap Y$ must contain both A and B. There is only one such pair, namely AB.

$$F = \{AB\}.$$

(v) The event $G = X \cup Y$ consists of those pairs containing either the letter A, or the letter B, or both: there are seven such pairs.

$$G = \{AB, AC, AD, AE, BC, BD, BE\}.$$

(vi) X' consists of those pairs not containing A.

$$X' = \{BC, BD, BE, CD, CE, DE\}.$$

(continued from page 5)

Thus from the previous discussion, the sample space for the experiment as specified is

$$S \times S \times S = \{(H, H, H), (H, H, T), (H, T, H), (H, T, T), (T, H, H),$$
$$(T, H, T), (T, T, H), (T, T, T)\}.$$

The following diagram is a tree diagram for this experiment.

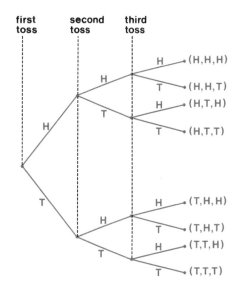

The number of elementary events in this sample space is

$$N(S) \times N(S) \times N(S) = 2 \times 2 \times 2 = 8.$$

Exercise 1

Exercise 1
(2 minutes)

Draw a tree diagram for the sample space for the experiment of throwing one die and tossing one coin simultaneously.

18.2.3 The Urn Model

Sometimes operations which at first glance appear quite different do in fact lead to identical outcomes. Thus whether we throw a die or draw a ball from six balls (each carrying a different number from 1 to 6) from an urn, the sample spaces for a single trial will be identical in the two cases. When situations which look different are identical in their essentials, this will manifest itself in the mathematics which describes the situations, and it may mean that a solution worked out in one case can immediately be applied to the other. Also, it means that we can work in terms of the situation which is the most convenient or helpful.

It often pays to work in terms of the situation in which balls are drawn from an urn. Not only can the various possible outcomes be envisaged clearly, but also we can introduce an important variation. For, if we have an experiment consisting of a sequence of trials, we can specify that each ball drawn from the urn is either

(i) replaced

or

(ii) not replaced

before the next ball is drawn. These types of selection are called selection with replacement and selection without replacement respectively.

If we choose to work with the corresponding urn situation rather than the actual situation, we say that the urn is a *model* of what actually happens. Hence we get the expression the urn model.

Selection with Replacement

An example of *selection with replacement* is provided by the next exercise.

Exercise 1

An urn contains 3 red balls, labelled R_1, R_2, R_3, and 2 white balls, labelled W_1, W_2. A ball is drawn and replaced, and then a ball is drawn a second time. The labels of the balls drawn are noted in *the order in which they are drawn*.

  (i) Write down an expression for the sample space corresponding to the trial described above, and draw the corresponding tree diagram.
 (ii) How many sample points does this sample space contain?
(iii) How many of these sample points contain (a) no white balls, (b) 1 white ball, (c) 2 white balls? ■

Selection without Replacement

Suppose a street contained 10 householders, and you were told to interview 3 of them chosen at random. How would you decide which to interview, and in which order? One thing you could do would be to use the urn model; that is, write the house numbers on balls, put the balls in an urn, and then draw out 3 of the balls. But, having drawn one house number, say no. 4, you would want to interview *different* people on the next two occasions. The obvious way of ensuring that you do get someone different is not to replace ball no. 4 in the urn; similarly, you do not replace the second ball. This operation is called *selection without replacement*.

Exercise 2

  (i) An urn contains 3 red balls and 2 white balls labelled as in Exercise 1. Two balls are drawn, but this time *without replacement*. If the order in which the two balls are selected is still material, draw the tree diagram corresponding to the selection procedure. How many sample points are there in the sample space?
 (ii) How many of these sample points contain (a) no white balls, (b) 1 white ball, (c) 2 white balls? ■

Solution 18.2.2.1

Solution **18.2.2.1**

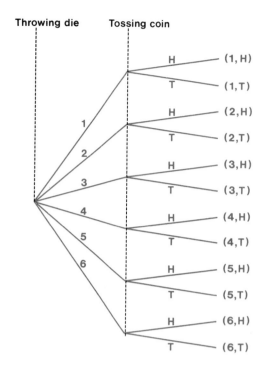

Solution 1 **Solution 1**

(i) The compound sample space is $S \times S$, where
$S = \{R_1, R_2, R_3, W_1, W_2\}$.
The corresponding tree diagram is:

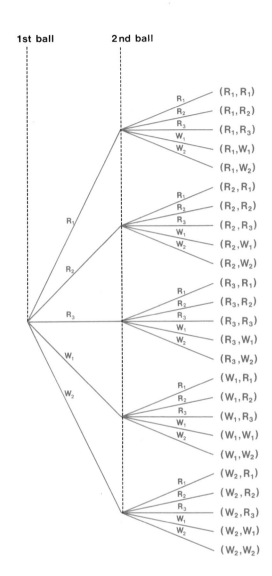

(ii) $5^2 = 25$.
(iii) (a) Number containing no white balls is 3^2 $\qquad = 9$
    (b) Number containing 1 white ball is $3 \times 2 + 2 \times 3 = 12$
    (c) Number containing 2 white balls is 2^2 $\qquad = \dfrac{4}{25}$

Solution 2 **Solution 2**

(i)

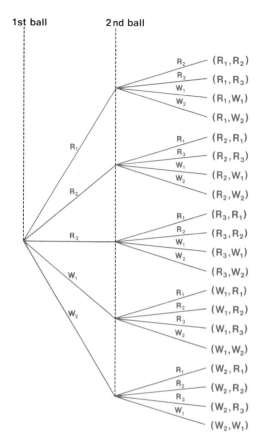

1st ball   2nd ball

The number of sample points is $5 \times 4 = 20$

(ii) (a) Number containing no white balls is $3 \times 2 \qquad = \quad 6$
    (b) Number containing 1 white ball is $3 \times 2 + 2 \times 3 = 12$
    (c) Number containing 2 white balls is $2 \times 1 \qquad = \quad \underline{\quad 2 \quad}$
    $$\underline{20}$$

18.2.4 Permutations and Combinations

In the last section we wondered how we could select 3 householders out of 10. In some probability situations (which we shall discuss later in this text) it is important to know how many ways there are of making such selections. If the order is material, and the 10 people are designated by the letters $A, B, C, D, E, F, G, H, I, J$, the various selections are:

$(A, B, C), (A, B, D), \ldots \qquad \ldots (A, B, J),$
$(B, A, C), (B, A, D), \ldots \qquad \ldots (B, A, J),$
$(A, C, B), (A, C, D), \ldots \qquad \ldots (A, C, J),$
$\ldots$
$(J, I, A), (J, I, B), \ldots \qquad \ldots (J, I, H).$

Ordered selections of r objects from a set of n objects are called permutations. Our list above consists of all permutations of 3 objects from 10.

An essential aspect of a permutation is that it is concerned with *order*. Thus the permutation (A, B, C) is *different* from the permutation (B, A, C). The number of permutations of r objects from n is denoted by $^{n}P_{r}$ (read as "nPr"). In the special case when $n = r$, we have $^{n}P_{n}$, the number of possible arrangements of n objects.

In probability, we are particularly concerned with the *value* of $^{n}P_{r}$. Later in the course (when we discuss *groups*), we shall be interested in the actual *ways* in which n elements can be rearranged.

Obviously we do not want to go through the chore of writing down every permutation, just to count up the value of ^nP_r, especially when there is a simple formula for calculating it. You will understand the calculation better, however, if you work the following exercise and try to obtain the formula for yourself.

Exercise 1

(i) Given the 5 letters A, B, C, D, E (considered as physical objects), how many ways are there of selecting a single letter?

(ii) A single letter, say A, having been selected and *removed*, how many ways are there of selecting a single letter from those remaining?

(iii) If the first letter selected and removed had been any letter (not necessarily A), would the number of ways of selecting the second letter still have been the same?

(iv) Complete the following sentences:
For each selection of the first letter, there are ... ways of selecting the second.
There are ... ways of selecting the first letter.
Therefore there are ... times ... ways of selecting an ordered pair of letters from the original 5.

(v) Pursuing the same ideas, write down in product form the number of permutations of 3 letters from 10.

(vi) Similarly, write down in product form the number of permutations of r objects from n.

(vii) How many permutations are there of n objects from n?

So far in this section we have been interested in *ordered* selections, but the ordering could be immaterial to us. For example, when deciding which 3 people to interview out of 10, we could arrange the order of interviewing to suit ourselves rather than be bound by the order in which the balls came out of the urn. In this case the only important issue settled by the procedure is the set of people chosen, *not* the order in which they are chosen. In other words, we are selecting a set of 3 objects from 10. In the general case, we may be interested in selecting a set of r objects from a set of n objects. It is usual to call a set of r objects selected from n objects a combination.

The situation is confused by football pools companies who use the word *permutation* for what mathematicians call a *combination*. What matters here is simply the set of teams you have selected in your treble chance entry, not the order in which you made up your mind or marked them in. If you go in for the so-called permutation schemes and you mark in 10 matches, you are really interested in the *combinations* of 8 matches from 10.

To illustrate the difference between permutations and combinations, we select two letters from the set $\{A, B, C\}$.

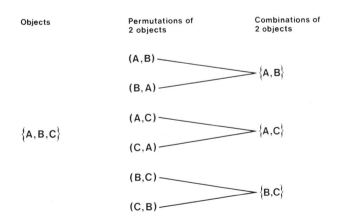

Objects	Permutations of 2 objects	Combinations of 2 objects

(continued on page 13)

Solution 1

(i) 5

(ii) 4

(iii) Yes

(iv) 4

5

5 times 4 = 20

(v) $10 \times 9 \times 8 = 720$

(vi) $n(n-1)(n-2) \times \cdots \times (n-r+1)$.

This answer can be expressed using factorials:

(See RB9)

$n(n-1)(n-2) \times \cdots \times (n-r+1)$

$$= \frac{n(n-1) \times \cdots \times (n-r+1)(n-r)(n-r-1) \times \cdots \times 3 \times 2 \times 1}{(n-r)(n-r-1) \times \cdots \times 3 \times 2 \times 1}$$

$$= \frac{n!}{(n-r)!}.$$

Hence we can now write

$$^nP_r = \frac{n!}{(n-r)!}$$

(vii) From the formula above,

$$^nP_n = \frac{n!}{0!} = \frac{n!}{1} = n!$$

Alternatively, we can select the first object in n ways, the second object in $(n-1)$ ways, ..., the nth object in 1 way, so

$$^nP_n = n(n-1) \times \cdots \times 2 \times 1 = n!$$

■

Notice here that we have used { } to denote combinations, because a combination of 2 objects is the same thing as a set of 2 objects; order is immaterial in both cases. But we have used ( ) to denote permutations, where order is material.

(continued from page 11)

The number of combinations of r objects from n is denoted by ^nC_r (read as "nCr"). We shall derive a formula for ^nC_r, just as we did for ^nP_r. Once again we shall do it step by step in an exercise.

Notation 2
* * *

Exercise 2

Exercise 2
(3 minutes)

(i) Draw a diagram (similar to the last diagram) showing the permutations and combinations of 3 objects from a set of 4 objects, say $\{A, B, C, D\}$.

(ii) Just by counting, how many combinations are there?

(iii) How many permutations can be obtained from each combination?

(iv) Complete the following sentences:

In the general case of n objects when each combination contains r objects, the number of permutations linked with each combination is ...

There are ^nC_r combinations, therefore there are ... permutations in all. From Exercise 1, the number of such permutations is

$$\frac{n!}{(n-r)!}.$$

Therefore ^nC_r is equal to ...

(v) Check that the formula derived in (iv) gives the right answer for the case where $n = 5, r = 2$.

(vi) The number of combinations of 2 objects from 5 objects equals the number of combinations of 3 objects from 5. Is this to be expected?

18.2.5 Summary

18.2.5

The following statements refer to the selection of r objects from n objects.

Summary
* * *

(i) There can be selection with replacement.

(ii) There can be selection without replacement.

(iii) Order within the selection can be material.

(iv) Order within the selection can be immaterial.

(v) In the "without replacement" case, ordered selections are called *permutations*, and unordered selections are called *combinations*.

(vi) The number of permutations of r objects from n is

$$^nP_r = \frac{n!}{(n-r)!}.$$

(vii) The number of combinations of r objects from n is

$$^nC_r = \binom{n}{r} = \frac{n!}{(n-r)!\,r!}.$$

Solution 18.2.4.2 Solution **18.2.4.2**

(i)

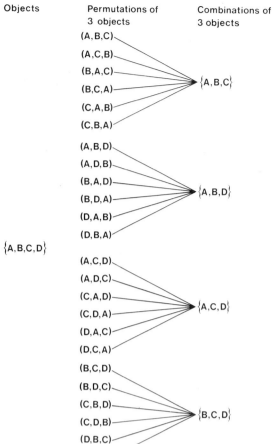

(ii) 4

(iii) $6 = 3!$

(iv) $^rP_r = r!$

$r! \times {}^nC_r$

$\dfrac{n!}{r!(n-r)!}$ which is equal to the binomial coefficient $\begin{pmatrix} n \\ r \end{pmatrix}$. (See RB9)

As the answer is $\begin{pmatrix} n \\ r \end{pmatrix}$, we shall drop the written notation ^nC_r completely. From now on the number of combinations of r objects from n will be written as $\begin{pmatrix} n \\ r \end{pmatrix}$. In speech we usually read $\begin{pmatrix} n \\ r \end{pmatrix}$ as "nCr".

(v) There are 10 combinations of two letters from $\{A, B, C, D, E\}$. They are

$$\{A, B\}, \{A, C\}, \{A, D\}, \{A, E\}, \{B, C\},$$

$$\{B, D\}, \{B, E\}, \{C, D\}, \{C, E\}, \{D, E\}.$$

$$\begin{pmatrix} 5 \\ 2 \end{pmatrix} = \frac{5!}{2!3!} = \frac{5 \times 4}{1 \times 2} \times \frac{3 \times 2 \times 1}{3!} = \frac{20}{2} = 10$$

(vi) Yes.

Instead of choosing two letters for selection, we could choose three letters for non-selection. The two processes are equivalent. Hence, in this case, the number of combinations of 2 objects must be equal to the number of combinations of 3 objects. In general, the number of combinations of r objects from n is equal to the number of combinations of $n - r$ objects from n. This is a common-sense way of seeing this result, though it follows immediately from the fact that

$$\begin{pmatrix} n \\ r \end{pmatrix} = \frac{n!}{r!(n-r)!} = \frac{n!}{(n-r)!r!} = \begin{pmatrix} n \\ n-r \end{pmatrix}$$

18.3 RULES OF PROBABILITY

18.3.0 Introduction

We know intuitively what we mean by *probability*; we have described it as relative frequency in the long run. It was when we tried to tighten this up into a formal definition that we ran into difficulties.

If it were not for those difficulties, we could have said exactly what we mean by probability, and in any simple case we could then have decided what *probability value* to attach to each sample point, and indeed to each event in the sample space. We shall now pursue the question strictly within the context of sample spaces containing a finite number of sample points. For obvious convenience, we suppose all such sample points to have non-zero probability, for if not we can delete them. From a set containing n elements we can form 2^n subsets. (Why?*) Thus, in a sample space of n points there are 2^n events; each must have a probability, and there are obviously relationships governing these 2^n probabilities. It is our job here to decide what those relationships are. We shall use our intuitive concepts associated with relative frequency to help frame the precise rules (axioms) of probability. We shall denote the probability of an event A by $P(A)$.

18.3.1 Rule 1

If A is any event, then, thinking of $P(A)$ as a relative frequency in the long run, we know that any relative frequency lies between 0 and 1 (inclusive), that is, in the interval $[0, 1]$. So, however we ascribe probabilities, we must arrange to have

$$0 \leqslant P(A) \leqslant 1.$$

If S is the complete sample space, then in a sequence of n trials the event S occurs every time, since every trial leads to some outcome and hence to a sample point belonging to S. Therefore the relative frequency of occurrences of the event S is

$$\frac{n}{n} = 1.$$

As we are associating probability with relative frequency, we must ascribe probabilities so that

$$P(S) = 1.$$

Hence we get two relationships which together form Rule 1. These relationships are

$$0 \leqslant P(A) \leqslant 1$$
$$P(S) = 1$$

18.3.2 Rule 2

If we know the probabilities of various subsets of the sample space, it is clearly an advantage if we can calculate the probabilities of various combinations of subsets. The two most basic combinations are union and intersection. First, we deal with union. The probability of $A \cup B$ will be the probability that *either* A happens *or* B happens *or* both. If A and B

---

* This result can be established using a proof by induction (see *Unit 17*). Note that a set with only one element has 2 subsets: itself and $\varnothing$.

are exclusive events (i.e. $A \cap B = \varnothing$) with probabilities $P(A)$ and $P(B)$ respectively, and in a sequence of n trials A occurs m_1 times and B occurs m_2 times, then the m_1 occasions are different from the m_2 occasions, since A and B never occur together. Therefore the number of occasions where *either A or B* occurs is $m_1 + m_2$.

Expressing this result in terms of relative frequencies we have:

the relative frequency of the occurrence of the event $A \cup B$ is $\dfrac{m_1 + m_2}{n}$,

where A and B are exclusive events.

Now

$$\frac{m_1 + m_2}{n} = \frac{m_1}{n} + \frac{m_2}{n}$$

$$= \text{relative frequency of } A + \text{relative frequency of } B.$$

We are associating probability with relative frequency, so however we finally assign probabilities, the ascribed probabilities should obey the rule:

if A and B are exclusive events, then
$P(A \cup B) = P(A) + P(B)$

Rule 2
* * *

That is, if A and B are *exclusive*, the probability of A or B is the probability of A + the probability of B.

Deductions

Simple though these rules are, we are now able to draw a number of conclusions:

(i) If A, B, C are mutually exclusive events (that is, no sample points belong to more than one of these sets), then the events A and $B \cup C$ are exclusive (see 18.2.1).

$$\therefore P(A \cup B \cup C) = P(A \cup (B \cup C)) \qquad (\cup \text{ is associative})$$

$$= P(A) + P(B \cup C) \qquad (\text{Rule 2})$$

$$= P(A) + P(B) + P(C) \qquad (\text{Rule 2}).$$

This result can obviously be extended to four and more mutually exclusive events. (A strict proof could be accomplished using mathematical induction: see *Unit 17, Logic II*.)

(ii) If the event A consists of the sample points

$$a_1, a_2, \ldots, a_n,$$

and the corresponding elementary events are

$$A_1, A_2, \ldots, A_n, (\text{i.e. } A_i = \{a_i\} \ (i = 1, 2, \ldots, n)),$$

then

$$A = A_1 \cup A_2 \cup \cdots \cup A_n,$$

where

$$A_1, A_2, \ldots, A_n$$

are mutually exclusive.

$$\therefore P(A) = P(A_1 \cup A_2 \cup \cdots \cup A_n)$$

so

$$P(A) = P(A_1) + P(A_2) + \cdots + P(A_n).$$

(iii) If S is composed of sample points $a_1, a_2, \ldots, a_n$ which all have the *same probability* p, and the corresponding elementary events are $A_1, A_2, \ldots, A_n$, then

$$1 = P(S) \qquad \text{(Rule 1)}$$

$$= P(A_1) + P(A_2) + \cdots + P(A_n) \quad \text{(see (ii))}$$

$$= np$$

$$\therefore p = \frac{1}{n}.$$

Hence if A is the event

$$A_1 \cup A_2 \cup \cdots \cup A_m,$$

then

$$P(A) = P(A_1) + P(A_2) + \cdots + P(A_m)$$

$$= mp$$

$$= \frac{m}{n}.$$

A quick way to find the probability of A in such cases as this is therefore to divide the number of sample points in A by the total number of sample points.

(iv) If A is any event in a sample space S, and A' is the event complementary to A, then A and A' are exclusive, and $S = A \cup A'$.

$$\therefore 1 = P(S) \qquad \text{(Rule 1)}$$

$$= P(A \cup A')$$

$$= P(A) + P(A') \qquad \text{(Rule 2)}$$

Therefore

$$P(A') = 1 - P(A). \qquad \qquad \text{* * *}$$

(v) We can now consider the case where $P(A) = 0$. We have

$$P(A) = 0 \underset{\text{(by iv)}}{\Leftrightarrow} P(A') = 1 \underset{\text{(Rule 1)}}{\Leftrightarrow} A' = S \Leftrightarrow A = \varnothing$$

That is, the probability of an event is zero if and only if the event is impossible (see the discussion at the top of page 3).   * * *

(vi) If A and B are any two events, we know that

$$A = A \cap S \qquad (A \subseteq S)$$

$$= A \cap (B \cup B') \qquad (B \cup B' = S)$$

$$= (A \cap B) \cup (A \cap B') \qquad \begin{array}{l} (\cap \text{ is distributive over } \cup \text{: see} \\ \textit{Unit 6}). \end{array}$$

But $A \cap B$ and $A \cap B'$ are necessarily exclusive (see diagram).

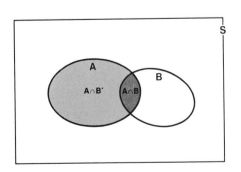

It follows that

$$P(A) = P(A \cap B) + P(A \cap B') \qquad \text{(Rule 2)}$$

$$\therefore \; P(A \cap B') = P(A) - P(A \cap B)$$

* * *

(vii) If B is a subset of A, we have

$$A \cap B = B \qquad\qquad\qquad (B \subseteq A)$$

so

$$P(B) = P(A \cap B)$$
$$= P(A) - P(A \cap B'), \qquad \text{(by (vi))}$$

so

$$P(B) \leqslant P(A) \qquad\qquad (P(A \cap B') \geqslant 0 \text{ by Rule 1})$$

(viii) We can think of $A \cup B$ as made up of two non-overlapping subsets (exclusive events), B and that part of A which is not in B, that is, $A \cap B'$ (see the previous diagram). Thus

$$A \cup B = (A \cap B') \cup B$$

and

$$P(A \cup B) = P((A \cap B') \cup B)$$
$$= P(A \cap B') + P(B) \qquad \text{(Rule 2)}$$

so

$$P(A \cup B) = P(A) + P(B) - P(A \cap B) \qquad \text{(by (vi))}.$$

* * *

This result is very important; it is an extension of Rule 2, since it applies to *any* two events A and B, exclusive or not. Notice that we have proved* it from the rules (axioms) of probability and the set axioms.

Applications

We are now in a position to tackle a number of problems. Solving a problem usually means finding the probability of some event, given the probabilities of the sample points. How we know or assess the values of these latter probabilities will be discussed in section 18.4.

Applications

Example 1

Example 1

Assuming that all numbers on a die have an equal probability of occurring, what is the probability of getting:
 (i) a prime number (1 is not a prime number)?
(ii) a composite number (one that has factors other than 1 and itself)?
(iii) Do these two probabilities sum to unity? ■

Solution of Example 1

 (i) (See deductions (ii) and (iii) in the text.) All sample points have the same probability. There are 6 sample points in all, so they each have probability $\frac{1}{6}$. 3 sample points 2, 3 and 5, correspond to prime numbers, therefore the probability of getting a prime is $\frac{1}{6} + \frac{1}{6} + \frac{1}{6} = \frac{1}{2}$.
(ii) Two of the outcomes are composite numbers, 4 and 6, so the probability of getting a composite number is $\frac{2}{6} = \frac{1}{3}$.
(iii) The sum of the two probabilities is

$$\tfrac{1}{2} + \tfrac{1}{3} = \tfrac{5}{6} \neq 1$$

because the number 1 is not in either of the events. (It is neither prime nor composite.) ■

---

* We have not set out the steps of the proof as strictly as *Unit 17* requires, but the main steps of the argument are there.

Exercise 1

(i) A coin is tossed twice. Assuming that all possible outcomes are equally likely, what is the probability of obtaining just one head out of the two tosses?

(ii) A die is thrown twice; assuming that all possible outcomes are equally likely, what is the probability of getting a total of 9?

(iii) In (ii), what is the probability of getting more than 9?

(iv) A pack of 52 playing cards is shuffled, and then the top two cards are drawn. Assuming that all pairs are equally likely, what is the probability that the pair consists of two aces?

(v) Alex tosses 2 pennies and Bob tosses 3 pennies. Assuming all possible outcomes are equally likely, what is the probability that Bob gets more heads than Alex?

(vi) In an aircraft the probability that the automatic landing device fails is 10^{-7}, and the probability that the fuel system fails seriously is also 10^{-7}. Can you give a useful upper bound for the probability that at least one of these contingencies occurs? ∎

Solution 1

Solution 1

(i) The sample space of all possible outcomes is

$$\{(H, H), (H, T), (T, H), (T, T)\},$$

and it is given that the sample points all have the same probability. Two of the sample points contain a single head; so the probability of a single head is $\frac{1}{2}$.

(ii) For two throws of a die, the number of sample points is 36. Four of these give a total of 9, namely $(3, 6), (4, 5), (5, 4), (6, 3)$, so the probability of getting a total of 9 is $\frac{4}{36} = \frac{1}{9}$.

(iii) Six sample points give a total greater than 9, namely $(5, 5), (6, 4), (4, 6)$, $(6, 5), (5, 6), (6, 6)$.

It follows that P (total > 9) $= \frac{6}{36} = \frac{1}{6}$.

(iv) The number of possible (ordered) pairs is 52×51. Of these, 4×3 consist of two aces, so that the probability that both cards drawn are aces is

$$\frac{4 \times 3}{52 \times 51} = \frac{1}{13 \times 17} = \frac{1}{221}.$$

(v) There are 4 possible outcomes for Alex, and for each of these Bob has 8 possible outcomes. (We are regarding (T, H) as different from (H, T) etc.) Therefore compounding the results for the experiment as a whole, there are 32 different outcomes, all equally likely. By examining these you can verify that Bob gets more heads than Alex with probability $\frac{1}{2}$. There is, however, an instructive short cut. Looked at in a different way, there are only two possible outcomes: either Alex gets fewer heads than Bob, or he gets fewer tails than Bob. These two outcomes are exhaustive and exclusive, that is, one and only one of them must occur; further, they are equally likely. By symmetry, they have equal probability, say p. Then $p + p = 1$, so $p = \frac{1}{2}$.

(vi) If A is the event of failure in the automatic landing device, and B is that of serious failure in the fuel system, the compound event that at least one failure occurs is $A \cup B$. Therefore, the required probability is

$$P(A \cup B) = P(A) + P(B) - P(A \cap B) \qquad \text{(deduction (viii) in the text)}$$

$$\leqslant P(A) + P(B) \qquad \text{(since } P(A \cap B) \geqslant 0)$$

$$= \frac{2}{10^7}.$$

Hence we have an upper bound for the probability of a failure somewhere.

By an extension of these arguments, if there are 10 possible kinds of failure, and each has probability of $\dfrac{1}{10^7}$ of occurring, the probability of a failure occurring somewhere is not greater than $\dfrac{10}{10^7} = \dfrac{1}{10^6}$.

We cannot get a closer estimate without carefully considering such terms as $P(A \cap B)$.

18.3.3 Conditional Probability

Rule 3

We have already worked out the probability of drawing two aces (without replacement) from a shuffled pack; we can, however, look at this trial in a different way.

Let A be the event of the first card being an ace, and B be the event of the second card being an ace. We are interested in the compound event $A \cap B$. Suppose the whole experiment is carried out N times, where N is large, and that in M_1 of these the first card is an ace, and in M_2 of these both cards are aces. Then clearly the M_2 cases are contained in the M_1 cases.

Now the number of times we get two aces is M_2, so the relative frequency of success is

$$\frac{M_2}{N} = \frac{M_1}{N} \times \frac{M_2}{M_1}.$$

$\dfrac{M_1}{N}$ is the relative frequency of drawing an ace for the first card; it will therefore be close to $P(A)$ when N is large. Now, out of the M_1 cases where we get an ace with the first card, we get an ace with the second card M_2 times, and so $\dfrac{M_2}{M_1}$ is the relative frequency of drawing an ace for the second card when we restrict ourselves to those occasions when an ace is produced for the first card. It will therefore be close to the probability of B when A has already occurred. We therefore want a notation for the probability of B, given that A has occurred; the standard notation is $P(B/A)$. This is called a conditional probability, the condition being of course that A has occurred. Allowing relative frequency, once again, to suggest our rules (axioms) for probability, we have

$$P(A \cap B) = P(A) \times P(B/A)*.$$

Looking at numerical values for our card example, when A has occurred there are only 51 cards left in the pack, and of these only 3 are aces, so that

$$P(B/A) = \tfrac{3}{51}.$$

The first thing to notice is that the value of this probability does depend on the occurrence of A. If the first card had not been an ace, there would still be 4 aces left in the pack; therefore the probability of B would be $\tfrac{4}{51}$. Secondly, conditional probability is a probability. If you were given a pack of cards which had lost one of its aces, and you drew one card, the straightforward probability of its being an ace is $\tfrac{3}{51}$. In other words, conditional probability is not a different kind of probability, but ordinary probability in a different or modified setting. In fact, all we have done is to change the sample space from an ordinary pack of 52 cards to a pack of 51 cards containing only three aces. In general, conditional probability can be viewed in this light; it is a midstream change in the choice of sample space.

Sometimes we shall use the value of $P(B/A)$ to obtain $P(A \cap B)$, and sometimes $P(A \cap B)$ to obtain $P(B/A)$; this is a matter of practical convenience for the problem under consideration.

---

* Our justification of this rule relies on the relation $\dfrac{M_2}{N} = \dfrac{M_1}{N} \times \dfrac{M_2}{M_1}$ and this breaks down if $M_1 = 0$ – which will be the case if $P(A) = 0$. In writing Rule 3 we therefore tacitly assume that $P(A) \neq 0$. We do not need a rule for $P(A \cap B)$ if $P(A) = 0$, since then we know that $P(A \cap B) = 0$; so nothing is lost.

Exercise 1

Rule 3 gives an expression for $P(A \cap B)$ in terms of conditional probabilities. Find the corresponding expression for $P(A \cap B \cap C)$. ∎

18.3.4 Applications

Having introduced the three rules of probability, we shall now consider them in some applications.

Example 1

Example 1

In the game of poker, five cards are dealt to each player from a well-shuffled pack. The order in which a player receives his cards is of no importance; what matters is the set of cards he receives. How many different five-card hands can be dealt from the pack? What is the probability of being dealt the royal spade flush?

∎

Solution of Example 1

The number of hands is the number of combinations of 5 objects from 52; we saw in section 18.2.4 that this is equal to

$$\binom{52}{5} = \frac{52!}{5! \times 47!} = \frac{52 \times 51 \times 50 \times 49 \times 48}{5 \times 4 \times 3 \times 2 \times 1}$$

$$= 2\,598\,960$$

$$\simeq 26 \times 10^5.$$

If the game is not taking place in a cowboy film, then we assume that any combination is as likely as any other, so that the required probability is approximately

$$\tfrac{1}{26} \times 10^{-5}.$$ ∎

Example 2

Example 2

An urn contains 5 white balls and 3 black balls. The urn is shaken, and then one ball is drawn and put on one side unobserved. What is the probability that a ball picked from the remaining 7 is white? ∎

Solution of Example 2

Denote by A the event of the unobserved ball being white; then A' is the event of its being black. Denote by B the event of the next ball being white; then

$$B = (A \cup A') \cap B = (A \cap B) \cup (A' \cap B),$$

where $A \cap B$ and $A' \cap B$ are exclusive.

Therefore, the probability that the second ball is white is

$$P(B) = P((A \cap B) \cup (A' \cap B))$$
$$= P(A \cap B) + P(A' \cap B)$$
$$= P(A) \times P(B/A) + P(A') \times P(B/A')$$
$$= \tfrac{5}{8} \times \tfrac{4}{7} + \tfrac{3}{8} \times \tfrac{5}{7}$$
$$= \frac{20 + 15}{56}$$
$$= \tfrac{5}{8}.$$

Exercise 1

Exercise 1
(5 minutes)

(i) Does the answer of $\tfrac{5}{8}$ in Example 2 surprise you? Can you think of any "short cut" argument?

(ii) *You may find it helpful to read the solution to part (i) before attempting this part.*

Four people sit down to a game of poker and each is dealt five cards from a well-shuffled pack. Player A receives 3, 4, 5, 6, 10 (suits ignored). He discards the 10, and is dealt another card in its place. What is the probability that this card completes a consecutive run of five in his hand?

Exercise 2

Exercise 2
(5 minutes)

In a batch of N manufactured items, it is known that α are defective. The defective items are not identifiable by sight. Of the N in the batch, n are drawn without replacement. It is known that $\alpha \geqslant n$ and $N - \alpha \geqslant n$.

(i) What is the probability that none of the n items is defective?

(ii) What is the probability that the n items contain exactly 1 which is defective?

(iii) What is the probability that the n items contain exactly r which are defective? ($r \leqslant n$)

(iv) Can you visualize a practical situation in which this information would be of importance?

Solution 18.3.3.1

Solution 18.3.3.1

$$P(A \cap B \cap C) = P((A \cap B) \cap C) \qquad (\cap \text{ is associative})$$

$$= P(A \cap B) \times P(C/(A \cap B)) \qquad \text{(Rule 3)}$$

$$= P(A) \times P(B/A) \times P(C/(A \cap B)) \qquad \text{(Rule 3)} \quad \blacksquare$$

Solution 1

Solution 1

(i) It should not surprise you (once you think about it). When you select 1 ball, you are equivalently rejecting 7. As long as you do not look at the colours during the process, it does not matter how you reject the 7. By selecting the second ball you reject 1 of the 7 by removing it from the urn, and you reject the other 6 without removing them from the urn. In effect therefore you are merely rejecting 7 of the 8 somehow, which is equivalent to selecting 1 of the 8. But for 1 ball out of 8, the probability of its being white is $\frac{5}{8}$. This example teaches us an important lesson. Although the second ball, when selected from the urn, is 1 out of 7, from the probability point of view it is as if it were 1 out of 8. It does not affect the probability if we merely carry out a separation process. It would affect the probability only if we gained information by looking at the colours of the balls separated off in this way.

(ii) After the initial deal, there are only $52 - 4 \times 5 = 32$ cards left in the pack. Therefore the extra card dealt to player A is physically 1 of 32. But as nothing is known about the 15 cards "separated off" to A's opponents, it is as if the extra card were 1 of $32 + 15 = 47$. These cards consist of four 2's, four 7's, and 39 other cards. Because of shuffling, each of these cards has probability $\frac{1}{47}$ of being dealt. Therefore the probability that A completes a run is the probability that he is dealt a 2 or a 7, which is $\frac{8}{47}$. $\qquad \blacksquare$

Solution 2

Solution 2

(i) Because the defective items are not identifiable by sight, we assume that all combinations of n items from the N are equally likely; the number of such combinations is $\binom{N}{n}$. The number of combinations containing no defective items is equal to the number of ways of drawing the n items from the $N - \alpha$ non-defective items, that is, $\binom{N - \alpha}{n}$.

Therefore the probability of getting no defective items is

$$\binom{N - \alpha}{n} \bigg/ \binom{N}{n}$$

$$= \frac{(N - \alpha)! \, n! \, (N - n)!}{n! \, (N - \alpha - n)! \, N!}$$

$$= \frac{(N - \alpha)! \, (N - n)!}{(N - \alpha - n)! \, N!}.$$

(ii) The 1 defective item can be selected in α ways. To each one of these ways there are $\binom{N - \alpha}{n - 1}$ ways of selecting the $n - 1$ non-defective items. Therefore the number of combinations containing just 1 defective item is $\alpha \binom{N - \alpha}{n - 1}$, so that the probability of getting just 1 defective item is

$$\alpha \binom{N - \alpha}{n - 1} \bigg/ \binom{N}{n}$$

$$= \frac{\alpha(N - \alpha)!\,n!(N - n)!}{(n - 1)!(N - \alpha - n + 1)!N!}$$

$$= \frac{\alpha n(N - \alpha)!(N - n)!}{(N - \alpha - n + 1)!N!}$$

(iii) The r defective items can be selected in $\binom{\alpha}{r}$ ways. The $n - r$ non-defective items can be selected in $\binom{N - \alpha}{n - r}$ ways.

Therefore, the probability of getting just r defective items is

$$\binom{\alpha}{r}\binom{N - \alpha}{n - r} \bigg/ \binom{N}{n}$$

$$= \frac{\alpha!(N - \alpha)!\,n!(N - n)!}{r!(\alpha - r)!(n - r)!(N - \alpha - n + r)!N!}$$

(iv) If you manufactured articles and sold them in batches of say 10 000 at a time, you might from experience expect there to be about 100 defective items in each batch. You could use the formulas we have just obtained to work out the probability of finding 1 defective item, or 2 defective items, etc., if you inspected a batch of, say, 40 items. By manipulation, you could get the probability of finding 1 or more defective items, or 2 or more defective items, etc. in the batch of 40. If in a particular case you found r defective items amongst the 40, and yet the theoretical probability of getting r or more was "very small", this would make you suspect that the batch of 10 000 in question had more than 100 defective items. This might lead you to throw the whole batch away, or at least to test it more thoroughly. (What is meant by "very small" would depend on the circumstances.) ∎

18.3.5 Statistical Independence

In section 18.3.3 we saw how the occurrence of one event A could affect the probability of the occurrence of another event B; the modified probability was called a *conditional probability* and denoted by $P(B/A)$. Let us look at this again.

Example 1

Example 1

We consider one throw of a fair* die, where

        A is the event $\{1, 2\}$;
        B is the event $\{1, 2, 3\}$;
        C is the event $\{2, 3, 4\}$.

(This example is slightly different from the one given in the television programme.)

The probability $P(B)$ is $\frac{1}{6} + \frac{1}{6} + \frac{1}{6} = \frac{1}{2}$.

Now Rule 3, which we deduced in section 18.3.3, tells us that

$$P(A \cap B) = P(A) \times P(B/A)$$

* A fair die (coin, etc.) is one for which all outcomes are equally likely.

In this case, $A \cap B = A$, so

$$P(B/A) = 1.$$

We can see this using common sense — if A occurs, then B must necessarily occur, so the probability of B occurring, *given* that A occurs, is 1. So $P(B) \neq P(B/A)$. The fact that A has occurred fundamentally affects the probability of B occurring.

Now looking at C, we have $P(C) = \frac{1}{2}$. What about $P(C/A)$? If A has occurred, the number thrown must have been a 1 or a 2. As the die is fair, these two possibilities are equally likely; so knowing that A has occurred, the probability of a 2 is $\frac{1}{2}$. But C will occur (as well as A) only if the number thrown is 2; so we have $P(C/A) = \frac{1}{2}$. This time

$$P(C/A) = P(C),$$

so the occurrence of A has no effect on the probability of C. It is as if C were *independent* of A.

Event A does not *cause* event B in the everyday physical sense of the word; similarly, the independence between A and C is not physical. ■

We now require a definition of statistical independence (as opposed to physical independence); we shall explore a little further before proceeding to a formal definition.

Main Text

Let A be an event such that $P(A) \neq 0$. We shall say that the event B is *statistically independent* of the event A if

$$P(B) = P(B/A);$$

Equation (1)

otherwise event B is *statistically dependent* on event A.

Situations exist in which there is a very close connection between physical dependence and statistical dependence.

Discussion

It is part of scientific philosophy that effects have causes. Doctors have therefore been searching for the agent physically causing cancer of the lung. But why were they looking for cancer agents in tobacco? Because there was already evidence of statistical dependence between heavy smoking and cancer of the lung. The evidence clearly showed that the conditional probability of contracting cancer of the lung, given that one smoked, was greater than the unconditional probability of contracting cancer of the lung (or, if you like, the conditional probability of contracting cancer of the lung if one did not smoke). Because of the apparent *statistical dependence*, the doctors felt that there was likely to be a *physical dependence* somewhere.

Put briefly, we can say:

statistical dependence implies physical dependence.

Let A and B be given events having non-zero probabilities, so that we have a right to expect them to occur (for example, "X smokes", "X has lung cancer").

If **p** is the proposition:

event B is statistically dependent on event A,

and **q** is the proposition:

event B is physically dependent on event A,

then **p** $\Rightarrow$ **q**. We know from *Unit 11, Logic I* that

if **p** $\Rightarrow$ **q**, then $\sim$**q** $\Rightarrow$ $\sim$**p**.

Also $\sim$**q** is the proposition:

event B is physically independent of event A,

and $\sim\mathbf{p}$ is the proposition:

> event B is statistically independent of event A.

So we have the proposition:

> *physical independence implies statistical independence.*

In a sense, we met statistical independence very early on in this text. In section 18.1 we talked of the probability of a 1 occurring in a random sequence always being the same "irrespective of the results so far". If we had had the notions available at the time, we could simply have said "the successive outcomes are statistically independent". To what extent are we justified in making this assumption? Can the throwing of a 6 with a fair die on one occasion affect the probability of a 6 on the next? It would seem that it cannot. But it is just possible that the die gets chipped or picks up some dirt as it lands, and ceases to be a fair die. In other words, there could be an unsuspected physical link all the time, which only statistical data would disclose. While no one need doubt that physical independence leads to statistical independence, what looks like physical independence may sometimes be illusory.

To return to probability relationships,* we are thinking of B as statistically independent of A if

$$P(B) = P(B/A).$$

Equation (1)

We can deduce a few simple facts from this statement and Rule 3, which states that:

$$P(A \cap B) = P(A) \times P(B/A).$$

Equation (2)

Since $\cap$ is commutative, we also have

$$P(A \cap B) = P(B) \times P(A/B).$$

Equation (3)

Combining Equations (2) and (3) gives

$$P(A) \times P(B/A) = P(B) \times P(A/B).$$

Substituting for $P(B/A)$ from Equation (1), we have

$$P(A) \times P(B) = P(B) \times P(A/B).$$

Cancelling by $P(B)\, (\neq 0)$ gives

$$P(A) = P(A/B).$$

Equation (4)

We have deduced Equation (4) from Equation (1), and similarly Equation (1) can be deduced from Equation (4), so we can write

$$P(B) = P(B/A) \Leftrightarrow P(A) = P(A/B).$$

We see that statistical independence is a symmetric relationship. If A is independent of B, then B is independent of A. This is scarcely surprising, but from the one-sided way in which we approached independence initially, it is a result which had to be proved.

Looking through the mathematics, we find that we have three important relationships expressing independence:

$$P(B) = P(B/A)$$

$$P(A) = P(A/B)$$

$$P(A \cap B) = P(A) \times P(B).$$

The last equation is obtained by combining either Equations (1) and (2) or Equations (3) and (4). Assuming that neither $P(A)$ nor $P(B)$ is zero, the truth of any one of these relationships establishes the truth of the other two.

* We assume here that $P(A) \neq 0$ and $P(B) \neq 0$.

All three are equivalent. It therefore does not matter logically which is adopted as the formal definition of *independence*. But as independence has been shown to be symmetrical between A and B, and as the last equation is the only one to exhibit this symmetry, we formally define statistical independence as follows:

Events A and B are statistically independent if

Definition 1
* * *

$$P(A \cap B) = P(A) \times P(B) \neq 0.$$

Events A and B, having non-zero probabilities, are statistically dependent if they are not statistically independent.

Exercise 1

Exercise 1
(3 minutes)

If A and B are statistically independent and B is a proper subset of the sample space S, show that A and B' are statistically independent. ■

Exercise 2

Exercise 2
(3 minutes)

(i) In an aircraft, the probability of failure in the automatic landing device is 10^{-7}, and that of bad failure in the fuel system is also 10^{-7}. Assuming that these failures are statistically independent, what is the probability of at least one of these failures occurring?

(ii) Would it be more dangerous if the failures were statistically dependent? (Examine the possibilities carefully.) ■

18.4 ASCRIBING PROBABILITIES

18.4

18.4.1 Relative Frequencies

18.4.1

If there is an experiment having a sample space S with elementary events $S_1, S_2, S_3, \ldots, S_N$ with probabilities $p_1, p_2, p_3, \ldots, p_N$ respectively, we can define a mapping:

Discussion
* * *

$$P : S_i \longmapsto p_i \qquad\qquad (i = 1, 2, \ldots, N)$$

So far we have not really considered the problem of deciding what value to ascribe to $p_i = P(S_i)$ $(i = 1, 2, \ldots, N)$.

Let us take the case of a coin. We might toss it 1000 times, getting 509 heads and 491 tails. Then if we want to use the experiment to attach a value to the probability of heads, p_1, we could take 0.509 or 0.51 (as being 0.509 rounded off to 2 decimal places) or 0.5 (as being 0.509 rounded off to 1 decimal place). At least, having carried out an experiment, we would obviously take some value related to our experimental results. The fact that we cannot decide on one single precise value need not worry us; nor need the fact that on a repetition of the experiment we are very likely to arrive at a different answer.

We know that $\pi = 3.14159265358979323\ldots$, yet we are perfectly happy to work with 3.14 in many cases.

Thus we see that our probability model in terms of sample spaces does not dissociate us from the relative frequency roots of probability — provided there is a relative frequency to which to turn. It is a way of formalizing the foundations of the subject, and of overcoming as far as possible the awkward circularity between probability and randomness.

18.4.2 Equally Likely Cases

While we must never maintain suppositions in defiance of experimental results, the fact that the relative frequency of heads is so close to $\frac{1}{2}$ in practice must make us wonder whether the $\frac{1}{2}$ was to be expected, and whether it can be relied on. At this point we might argue that a coin is a symmetrical thing: whatever can be said about the head side (apart from the design) can be said about the tail side. From this point of view, the probability of a head, however probability is defined, would seem to be equal to the probability of a tail. But as the two outcomes are exhaustive as well as exclusive, the probability of each must be $\frac{1}{2}$.

Exactly the same argument applies to a die, where there is symmetry between the 6 faces (design apart). Admittedly design differences mean the symmetry is incomplete; it is not unreasonable to assume, however, that differences of design have no effect — or if you prefer, negligible effect — on the dynamical behaviour of the die when thrown.

If we have an urn containing say 8 balls, then once again there is a basic symmetry in the situation, so that the probability of selecting any one can be expected to be equal to the probability of selecting any other.

18.4.3 Definitions of Probability and Randomness

Probability

In examining the outcomes of physical trials, we have used concepts such as sets, sample spaces and numbers. We have associated sample points with possible outcomes, and subsets of the sample space with events. In short, we have set up a mathematical model of the physical situation. In this model we then assigned numbers as images to the various subsets in such a way as to satisfy the rules of probability given in section 18.3. We now define such numbers to be probabilities.

If there is a function P, with domain the set of all events of a sample space, S, and codomain R, such that the images under P obey the rules of probability, then these images are the probabilities of the corresponding events.

This is very much a *mathematical definition*. It applies to the probability model rather than to the physical situation being modelled. If the physical situation conforms to the model, we have an adequate concept of physical probability; the trouble is that the model may not be accurate. How accurate it is we can only tell by attempting to measure the physical probability, and we do this by looking at relative frequencies (assuming of course that the trials are repeatable). We can carry out a crude measurement (by taking only a few trials) just as we can make a crude measurement of a length. We can get a more refined measurement by taking a large number of trials, just as we can make a more refined measurement of a length. But neither with the probability nor with the distance will we attain complete accuracy. In any case, how the relative frequencies are interpreted, and what conclusions we draw are questions taking us into the realm of *statistics*.

In a sense, to settle for a mathematical definition looks like second best. The trouble, in the view of most statisticians, is that any more direct form of physical probability just cannot be defined. In other words, it is a matter of having a mathematical definition or no formal definition at all.

(*continued on page 30*)

Solution 18.3.5.1

Since A and B are statistically independent, $P(A) \neq 0$ and $P(B) \neq 0$. Using the results obtained in section 18.3.2, we have:

$$
\begin{aligned}
P(A \cap B') &= P(A) - P(A \cap B) && \text{(deduction (vi))}\\
&= P(A) - P(A) \times P(B) && \text{(definition of}\\
& && \text{independence)}\\
&= P(A) \times (1 - P(B))\\
&= P(A) \times P(B') && \text{(deduction (iv))}\\
&\neq 0 && (B \subset S \Rightarrow B' \neq \varnothing)
\end{aligned}
$$

Hence A and B' are independent. ■

Solution 18.3.5.2

(i) If A is the event of bad failure in the fuel system, and B is the event of failure in the automatic landing device, then the event of getting at least one failure is $A \cup B$, and the probability is given by

$$
\begin{aligned}
P(A \cup B) &= P(A) + P(B) - P(A \cap B) && \text{(deduction (viii))}\\
&= P(A) + P(B) - P(A) \times P(B) && \text{(definition of}\\
& && \text{independence)}\\
&= 2 \times 10^{-7} - 10^{-14}
\end{aligned}
$$

which is less than 2×10^{-7}.

(ii) We cannot say. If A and B never occur together (if they are exclusive), then $P(A \cap B) = 0$. In this case, $P(A \cup B)$ is greater than in (i), showing that the aircraft is more dangerous (but not by much). If A and B always occur together, then $P(A \cap B) = P(A) = 10^{-7}$, so that the aircraft is only "half as dangerous". ■

(*continued from page 29*)

Randomness

Random is a word which we first applied to sequences of 0's and 1's. The sequences in question were the outcomes of trials, and it was observed that they were

(i) patternless — which is a collective property;
(ii) unpredictable in their terms — which is an individual property.

These sequences were called *random*, though this was not offered as a formal definition, for reasons already given. It was then observed that for such a sequence of 0's and 1's, the relative frequency of 1's behaved as if it were tending to a limiting value.

Now that we have a definition of probability, we can have a corresponding definition of randomness — as envisaged in section 18.0.

Normally we begin with a trial, define the sample space S in terms of the possible outcomes, and then ascribe probabilities to the events in S. Conversely, we could begin with a set S in which the elementary events had suitable numbers attached to them (*suitable* here means *obeying the rules of probability*). Then we could enquire whether there were any trial having S as sample space and the attached numbers as probabilities of the corresponding elementary events. If so, we shall say that the trial is random. We shall say also (though rather more loosely) that the outcome of a random trial is itself random.

(i) Notice that this definition agrees closely with popular conception; for example, if you were asked what it meant to "choose any one of the ten digits randomly", you would reply that in effect it meant having a trial in which on any occasion

(a) the outcome was one of the 10 digits

and

(b) the possible outcomes were equally likely.

In other words, you are specifying your sample space first, attaching the number $\frac{1}{10}$ to each sample point, and then calling a trial *random* if it has this space as sample space, and the attached numbers as probabilities.

The one discordant note is that this popular concept implies that probabilities must be equal for randomness to apply. This is not the case, as we shall see later in this section.

(ii) If a trial has sample space S with certain ascribed probabilities, then that trial is random relative to S and those probabilities.

(iii) Randomness is defined relative to a space S and attached numbers.

(This is no more peculiar than defining probability in terms of a trial and its outcomes.)

(iv) A *random sequence* is defined as a sequence of random outcomes.

Let us look at a simple case. If we take a set S containing two elements,

a_0 with attached value $\frac{1}{2}$, and
a_1 with attached value $\frac{1}{2}$,

then a random trial exists, namely the tossing of a fair penny (if we associate heads with a_0 and tails with a_1, for example).

If the points a_0, a_1 have attached values $\frac{3}{5}$, $\frac{2}{5}$, a random trial may exist in terms of a penny which is sufficiently biassed; alternatively, we can perform the trial of drawing balls from an urn as follows.

If there are five balls $b_1, b_2, \ldots, b_5$ in an urn, and we associate b_1 and b_2 with tails, and b_3, b_4 and b_5 with heads, we have the equivalent of a biassed penny for producing probabilities:

$P(\text{heads}) = \frac{3}{5}$
$P(\text{tails}) = \frac{2}{5}$.

Alternatively, we may suppose that 3 balls are white and 2 black, and then the stated probabilities correspond to the probability of drawing a white ball and a black ball respectively.

Choosing Randomly

Given a number of possible outcomes $a_1, a_2, a_3 \ldots$, we sometimes have to choose between them. Any choice we make is a kind of trial, and we shall have chosen randomly if the trial is random. But the definition of a random trial depends not only on outcomes $a_1, a_2, a_3, \ldots$, but also on attached numbers which (with foresight) may be represented by $P(a_1)$, $P(a_2)$, $P(a_3) \ldots$

Until these "probabilities" are specified, randomness is undefined in this situation.

When the man in the street asks you to choose randomly without specifying probabilities, implicitly he intends the probabilities to be equal. To his mind, events which are not equally likely are not random. But ask him whether in his view you get a random result if you throw two dice and then add the scores together. Ask him whether the guessing of suits in your card experiment (*Unit 16, Probability and Statistics I*) was random.

In our view, the card experiment is random relative to the probabilities

$$P(\text{correct guess}) = \tfrac{1}{4}$$
$$P(\text{incorrect guess}) = \tfrac{3}{4}$$

only if the relative frequency of correct guesses in the long run is $\tfrac{1}{4}$. If out of 500 guesses your friend gets say 130 right, you will say "Well that was just chance" (i.e. randomness). On the other hand, if he gets 250 right (note here that rightness and wrongness are equally likely on any single guess), you will say "That was no chance effect; that man has psychic powers".* It is clear that in this case you quite naturally examine randomness against specified probabilities which are not all equal. Until you had worked out the probabilities of $\tfrac{1}{4}$ and $\tfrac{3}{4}$ you would have been in no position to examine the issue of randomness at all.

Exercise 1

Exercise 1
(3 minutes)

Place 16 identical marks in a square as shown:

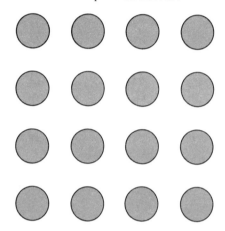

and invite different friends to "select one at random". (You intend the selection to be fair as well as random, and your friends will interpret your instruction in this manner anyway. But if you draw attention to the fact, you will spoil the experiment. It is therefore better to carry out this experiment before you open up the discussion referred to earlier in this section.) Note down how many choose one of the centre marks and how many choose one of the corner ones.

(i) Ought the two numbers to be about the same?
(ii) If they are not, can you suggest any reason why not?   ■

As it is so difficult to make random decisions, we fall back on physical devices such as ERNIE, balls in an urn, dice, etc. If we want to decide fairly and randomly between the ten digits, we could put ten identical balls into an urn and then choose one. Not having ten identical balls or an urn, we could look up the results of someone who did have them or their equivalent. These effectively are what tables of random numbers are; random numbers between 0 and 9 are the recorded outcomes of a random trial having as outcomes the ten digits all with equal probability.

Discussion
* *

Pretty well all the time we want random selection processes to be equiprobable; but not always so. If you manufactured gaming devices, you might want some outcomes to be more probable than others even though they *looked* equal.

You might want to give preference to children in some games involving chance, and you do this by arranging handicaps accordingly. In both these cases you want the outcomes (that is, winning or losing) to be settled by chance; you are merely aiming at certain unequal probabilities.

* Surely *your* friend wouldn't cheat?

Finally, let us see how random sequences as defined in this unit compare with random sequences as discussed in *Unit 16, Probability and Statistics I.* If we have a random sequence of 0's and 1's relative to probabilities p and $(1 - p)$, then each term will be unpredictable; also there will be an absence of regular pattern, for with a regular pattern the terms become predictable in time.

Thus our definition of random sequence fits in with our earlier notions. On the other hand, sequences earlier called random would not be recognized as random now, if only because no probabilities are specified. Whether there exist probabilities with respect to which any particular patternless unpredictable sequence is random is hard to say; the words *patternless* and *unpredictable* are so vague and negative that we have nothing to get a grip on.

18.4.4 Subjective Probability

So far we have been considering the probability of an event occurring in a trial which is either repeated or repeatable. The value of the probability is then associated with an actual or notional relative frequency. There can however be situations of uncertainty where the experiment is not repeatable, and yet where one has more confidence in one outcome than another, or possibly equal confidence. If numbers can be ascribed to the various outcomes representing one's various degrees of confidence, and these numbers satisfy the rules of probability, then they *are* probabilities (i.e. they satisfy the mathematical definition of probability). As they are arrived at personally, they are called subjective probabilities.

Each person estimates his own subjective probabilities on a basis of his general background experience. If two people differ in their subjective probabilities, the only thing to do is to obtain more information so as to try to settle the matter. This extra information adds to the experience of each person, and causes each to change his value of the probability (just as events might change an unconditional probability into a conditional probability with a different value). If both people are reasonable, we would expect their subjective probabilities to approach each other as the amount of information is increased. (This is presumably the philosophy behind "form books" for horse racing — by providing more information they assist the punter to judge the bookmaker's assessment of the probability of a horse winning a race.)

Having said all this, it does not follow that each person has a precise numerical subjective probability for some event E. Feelings are necessarily vague, and all one might be able to say is "My subjective probability that E occurs could lie anywhere between $\frac{1}{4}$ and $\frac{1}{2}$". In this situation he might be prepared to take a figure of $\frac{3}{8}$ for the sake of argument. But at least it is a start, and a crude estimate like this can be refined by additional information.

Subjective probability was introduced as something which could cover a "vacuum" when a trial was not repeatable. It is not restricted to this case, however. People instinctively (if tacitly) assume that the probability of a head is $\frac{1}{2}$ before they have tossed some new coin even once. This $\frac{1}{2}$ is their subjective probability. Because they have a subjective probability to start with does not mean they are relieved of the responsibility of tossing the coin. If the heads and tails come out with virtually the same frequency, this will reinforce the subjective value of $\frac{1}{2}$. If not, it should modify it.

Solution 18.4.3.1

(i) A choice which was fair and random would select a centre mark with probability $\frac{4}{16} = \frac{1}{4}$. Similarly for a corner mark. Therefore the numbers of centre marks and corner marks selected should be about the same.

(ii) If they are not the same, the usual reason is the psychological one that people feel corner positions to be particular or special, and hence "non-random". There is therefore an inbuilt bias to choose a more "general" centre position.

18.5 SUMMARY

This has completed the purpose of this unit, namely to lay down an intuitive foundation for probability theory. Now that we have that foundation, we can develop the subject of probability, solve probability problems, and launch ourselves into statistics. We have argued exactly in terms of sample spaces, and produced mathematical definitions of probability and of randomness. You should realize, however, that one could be even more exact and theoretical, though such extra rigour is seldom brought in except to cope with continuous situations, and not always then.

To achieve our ends, we have in some ways had to invert the natural order of things. Thus we have introduced rules of probability and even justified them before we gave our definitions of probability and randomness. Finally, we appeared to return to first principles by going into the ascription of probabilities and into notions of subjective probability. It is no new thing in mathematics, however, for a logical presentation of a subject to invert the natural order (as we saw in *Unit 17, Logic II*) especially where there are conceptual difficulties such as exist in probability.

But do not let all this axiomatic inversion disturb you. If you start to apply the subject to any extent, you will soon cease to worry over the axioms. *Unit 18* will still be of vital concern to you however, because it contains the pragmatic starting point of all applications, namely the rules of probability, together with concepts of independence. These are ideas which will never cease to be of use.

Postscript

"In the fell clutch of circumstance,
I have not winced nor cried aloud:
Under the bludgeonings of chance
My head is bloody, but unbowed."

William Ernest Henley
Echoes, iv. Invictus

Unit No.		Title of Text
1		Functions
2		Errors and Accuracy
3		Operations and Morphisms
4		Finite Differences
5	NO TEXT	
6		Inequalities
7		Sequences and Limits I
8		Computing I
9		Integration I
10	NO TEXT	
11		Logic I — Boolean Algebra
12		Differentiation I
13		Integration II
14		Sequences and Limits II
15		Differentiation II
16		Probability and Statistics I
17		Logic II — Proof
18		Probability and Statistics II
19		Relations
20		Computing II
21		Probability and Statistics III
22		Linear Algebra I
23		Linear Algebra II
24		Differential Equations I
25	NO TEXT	
26		Linear Algebra III
27		Complex Numbers I
28		Linear Algebra IV
29		Complex Numbers II
30		Groups I
31		Differential Equations II
32	NO TEXT	
33		Groups II
34		Number Systems
35		Topology
36		Mathematical Structures